**J.0. Iwouno**
**V.S. Igwe**

# Ethyl carbamate prevalence in spirits

J.0. Iwouno
V.S. Igwe

# Ethyl carbamate prevalence in spirits

LAP LAMBERT Academic Publishing

## Impressum / Imprint

Bibliografische Information der Deutschen Nationalbibliothek: Die Deutsche Nationalbibliothek verzeichnet diese Publikation in der Deutschen Nationalbibliografie; detaillierte bibliografische Daten sind im Internet über http://dnb.d-nb.de abrufbar.
Alle in diesem Buch genannten Marken und Produktnamen unterliegen warenzeichen-, marken- oder patentrechtlichem Schutz bzw. sind Warenzeichen oder eingetragene Warenzeichen der jeweiligen Inhaber. Die Wiedergabe von Marken, Produktnamen, Gebrauchsnamen, Handelsnamen, Warenbezeichnungen u.s.w. in diesem Werk berechtigt auch ohne besondere Kennzeichnung nicht zu der Annahme, dass solche Namen im Sinne der Warenzeichen- und Markenschutzgesetzgebung als frei zu betrachten wären und daher von jedermann benutzt werden dürften.

Bibliographic information published by the Deutsche Nationalbibliothek: The Deutsche Nationalbibliothek lists this publication in the Deutsche Nationalbibliografie; detailed bibliographic data are available in the Internet at http://dnb.d-nb.de.
Any brand names and product names mentioned in this book are subject to trademark, brand or patent protection and are trademarks or registered trademarks of their respective holders. The use of brand names, product names, common names, trade names, product descriptions etc. even without a particular marking in this work is in no way to be construed to mean that such names may be regarded as unrestricted in respect of trademark and brand protection legislation and could thus be used by anyone.

Coverbild / Cover image: www.ingimage.com

Verlag / Publisher:
LAP LAMBERT Academic Publishing
ist ein Imprint der / is a trademark of
OmniScriptum GmbH & Co. KG
Heinrich-Böcking-Str. 6-8, 66121 Saarbrücken, Deutschland / Germany
Email: info@lap-publishing.com

Herstellung: siehe letzte Seite /
Printed at: see last page
**ISBN: 978-3-659-35365-9**

# TABLE OF CONTENTS

**CHAPTER THREE**

**CHAPTER FOUR**

**CHAPTER FIVE**

## LIST OF TABLES

## 1.0 INTRODUCTION

Ethyl carbamate or Urethane (CAS 51-79-6) is an ethyl ester of carbamic acid. It can be found in fermented food and beverages like spirits, wine, beer, bread, Soy sauce and yoghurt; thus a by product of fermentation which occurs in fermented beverages (Conacher and page, 1986; Dennis *et al.*, 1989; Battaglia *et al.*, 1990a; Schlatter and Lutz, 1990; Zimmerli and Schlatter, 1991a; Sen *et al.*, 1992; Sen *et al.*, 1993; Benson and Beland, 1997; Kim *et al.*, 2000). The chemistry of action of ethyl carbamate are illustrated in figure 1:

$$R - C \equiv N \xrightarrow{\text{Enzymatic reaction}} H - C \equiv N \xrightarrow{\text{Oxidation}} HO - C \equiv N$$

Cyanogenic glucoside    Hydrocyanic acid    Hydrogen cyanate

Urea $+$ ethanol $\longrightarrow$ Ethyl carbamate

**Figure 1**: The formation of ethyl carbamate from ethanol and hydrogen cyanate (top) or from ethanol and urea (bottom).

Urethane is a white crystalline or granular powder and soluble in water (Anon, 1974a). One of the sources of urethane is by hydrolyzing ethanol to carbomyl phosphate and further to phosphoric acid and then urethane (Ough, 1976a). According to (Finar, 1951) it is an ester of carbamic acid, and that urea in ethanol at elevated temperature produces urethane. It was reported that cyanides are precursors of ethyl carbamate (Aylott *et al.*, 1987; Mackenzie *et al.*, 1990a; Mc Gill and Morley, 1990a). There are a number of precursors in food and beverages that can form ethyl carbamate including hydrocyanic acid, urea, citrulline, cyanogenic glycosides and other N-carbamyl compounds.

The potential of beer to develop ethyl carbamate from precursors is dependent upon the cereal source such as barley, rice, sorghum, maize and millet used in brewing the beverage (Cook *et al.*, 1990).

When fermented beverages are distilled such as in fruit brandies, the levels of urethane were found to increase in the distillate (Harry *et al.*, 1989a; Wardlaw and Insel, 1996a).

Urethane has been implicated as a chemical carcinogen and this has led to concern in recent years since it occurs in trace levels (less than 10ppb) in most alcoholic beverages and in other food items where fermentation is an integral part of the production process (Aylott *et al.*, 1990a, Conacher and Page, 1986b). Prolonged exposure of human to high concentration of ethyl carbamate may be hazardous to some body organs such as kidney, liver and could cause cancer, leukemia, and malignant tumor. The symptoms associated with high level of ethyl carbonate toxicity are: vomiting, difficulty in hearing, loss of vision and nausea (Anon, 1974b) .

However, currently there are no harmonized maximum levels for ethyl carbamate provided by Codex Alimentarius, the European Union, nor the countries under study.

Cyanic acid reacts with ethanol to form ethyl carbamate (Harry *et al.*, 1989b; Mackenize *et al.*, 1990b and Mc Gill *et al.*, 1990b). (Aylott *et al.*,1990b) reported on the cyanide content and the corresponding ethyl carbamate content in an alcoholic beverage [5% V/V] as shown:

Cyanide (mg/100mg): 0.50, 3.75, 4.50, 13.30 and ethyl carbamate (mg/100g): 1.00, 7.50, 1.25, 1.75, 2.25 respective.

The International Agency for Research on cancer (IARC) recently upgraded it's classification of ethyl carbamate to (Group 2A) "probably carcinogenic to humans". In 2007 (IARC, 2007) and also as "possibly carcinogenic to humans" (Group 2B). This reflects increasing evidence on the significant similarities between rodents and humans regarding the metabolic pathways of the activation of ethyl carbamate, whereby the formation of proximate DNA-reactive carcinogens, hypothesized to play a major role in ethyl carbamte-induced carcinogens in rodent cells, Is also likely to occur in humans. For example an ethyl carbamate exposure of 60ng/kg bw/day was estimated for spirit consumers in Europe resulting in a MOE (margin of exposure) of 5,000. Comparatively, the ethyl carbamate intake from the consumption of beverages under study does not differ from that in other countries like Mexico where the incidence of cirrhosis of the liver is lower; therefore, it seems unlikely that ethyl carbamate is a contributing factor. The ethyl carbamate exposure due to alcoholic beverages in the ng-range was also considerably lower than dosages (2.5 – 7.6g/day) known to lead to liver damage, based on instance when ethyl carbamate was used as medicine. Therefore the aim of this present work was to determine the prevalence of ethyl carbamate in ethyl alcohol from different material origin of production. It will also provide the impetus and framework for more researchers in this area.

# CHAPTER TWO

## 2.0 LITERATURE REVIEW

Urethane (ethyl carbamate) has long been considered a desirable compound for providing anesthesia during physiological experiments (Green, 1982a). However, in addition to the notable anesthetic properties, several adverse traits have been recognized when urethane is administered to animals and humans. These traits have been known for many years, yet they are relatively appreciated by many investigators. Therefore this material is aimed at reviewing the development, use, and adverse properties of urethane. Special consideration will be directed towards reviewing data that assists in handling of urethane (WHO, 1974a).

### 2.1 History

In 1939, a study was conducted on wheat seedlings using a variety of carbamate analogue. In an attempt to identify plant growth-promotants and growth inhibitors (Templeton and Sexton, 1945a). The data obtained from the study suggested that, depending on the level of exposure to ethyl phenyl carbamate (one of many ethyl carbamate analogues). Wheat seedlings would develop one or more of the following abnormalities: bulbous hypertrophy of the seed leaf or root tip, stunted growth, arrested growth at the seed-leaf stage, thick bulbous stem formation, or plant death (Templeton and Sexton, 1945b; Haddow and Sexton, 1946a). Other investigators confirmed these findings and observed similar results in winter rye, barley and Oat plants (Haddow and Sexton, 1946b). These investigators also noted that by varying the level of plant exposure to ethyl phenyl carbamate, they could inhibit weed growth without affecting the growth of desirable plants (Haddow and Sexton, 1946c). Some investigators, believing that the carbamates would contain several important and inexpensive herbicides, evaluated approximately 50 related aryl carbamates and thiocarbamates for their influence on plant growth and development (Templeton and Sexton, 1945c). At the end of this project, urethane proved to be one of the few analogues that failed to influence plant growth or development (Templeton and Sexton, 1945d; Haddow and Sexton, 1946d). Both ethyl phenylcarbamate and isopropyl phenylcarbamate, two of urethane's analogs, were known to retard the growth of the walker rat carcinoma 256.

Therefore urethane was further investigated as a potential agent against this tumor (Haddow and Sexton, 1946e). The action of urethane against the walker rat carcinoma was accompanied by profound modifications in the histological structure of the tumor which were suggestive of remission (Haddow and Sexton, 1946f). Intrigued with the thought that a compound as readily available and structurally simple as urethane would be of antineoplstic value, urethane was quickly introduced into the chemical setting for evaluation in human cancer patients (Haddow and Sexton, 1946g; Paterson *et al.*, 1946a). Shortly thereafter, urethane was administered to several human patients with advanced, inoperable, and otherwise intractable mammary carcinomas (Haddow and Sexton, 1946h; Paterson *et al.*, 1946b). Unfortunately the results of the initial clinical trials were negative (Haddow and Sexton, 1946i). Nevertheless, the investigators elected to pursue an extended course of therapy with these patients, a decision which provided to be fortuitous: chronic urethane therapy resulted in leucopenia

(Haddow and Sexton, 1946j; Paterson *et al.,* 1946c). Clinical trials on urethane were quickly extended to include individuals with leukemia and other lymphadenopathies and the results were encouraging (Haddow and Sexton, 1946k; Paterson *et al,* 1946d). Urethane has since been used in humans (total dose of 2-6g/day) as a chemotherapeutic agent against chronic granulocytic leukemia; multiple myeloma (Plasma-cell myeloma) and to a lesser extent, chronic lymphocytic leukemia (Haddow and Sexton, 1946l; Di palma, 1965). Urethane has also been used in humans as a hypnotic agent as an adjunct to sulphonamide therapy, as a component of sclerosing agent (with quinine) for varicose veins, and as a topical bactericide (WHO, 1974b). Currently there are several drugs that demonstrate higher specificity than urethane for treating these diseases in humans.

In humans, chronic urethane therapy results in leucopenia, sedation, nausea vomiting and hepatic necrosis (Hirschback *et al,* 1974a; Hoover, 1970a). The primary use of urethane today is as an anesthetic agent for laboratory animals (Hoover, 1970b; Jones *et al,* 1977a).

## 2.2 Occurrence in Beverages and Food

The discovery of the widespread of ethyl carbamate in alcoholic beverages first occurred during the mid 1980s. In 1987, the U.S centre for science in the public interest published _tainted Booze the consumers guide to urethane in alcoholic beverages_ to raise public awareness of this issue. Studies have shown that most, if not all, yeast-fermented alcoholic beverages contain traces of ethyl carbamate (15ppb to 12 ppm) other foods and beverages prepared by means of fermentation also contain ethyl carbamate for example bread has been found to contain 2ppb; as much as 20ppb has been found in some samples of soy sauce. Amounts of both ethyl carbamate and methyl carbamate have been found in wines sake, beer, brandy, whisky and other fermented alcoholic beverages.

Studies in Korea (2000) and Hong Kong (2009) out time the extent of the accumulative exposure to ethyl carbamate in daily life. Fermented foods such as soy sauce, kimchi, soybean paste, breads, rolls, buns, crackers and bean curd, along with wine, sake and plum wine, were found to be the food with the highest ethyl carbamate levels in traditional Asian diets.

In 2005, the JECFA (joint FAO/WHO Expert Committee on Food Addictives) risk assessment evaluation of ethyl carbamate concluded that the MOE intake of ethyl carbamate from daily food ad alcoholic beverages combined is of concern and mitigation measures to reduce ethyl carbamate in some alcoholic beverages should continue. Thus of the 6,376 sample results reported to the JECFA, the 372 food samples showed mean values from non-detected to 16µg/kg with an overall maximum of 84µg/kg in a soy sauce product. Among the 6,004 samples of alcoholic beverages there were some very high values of up to 6,131µg/kg reported in the group of cordials, liqueurs and brandies, presumably confirmed to brandies. The highest mean of 122µ/kg was reported in sake (FAO/WHO, 2006). It should be noted that some food and beverages groups contained very few samples as shown in table 2.1;

**Table 2.1: Concentrations of ethyl carbamate in food and beverages (FAO/WHO, 2006)**

| Product | Country | NO. of Samples | Mean (µg/kg) | Range (µg/kg) |
|---|---|---|---|---|
| **Alcoholic beverages** | | | | |
| Wine | Various | 5431 | 4-10 | ND-61 |
| Fortified wine | Various | 140 | 32-41 | ND-262 |
| Whisky | Various | 235 | 29-32 | ND-239 |
| Cordial, liqueur, brandy | Various | 14-31 | 37-64[a] | ND-243 |
| Sake | Japan | 92 | 73-122 | ND-202 |
| Beer | Various | 62 | ND[b]-1 | ND-5 |
| **Food** | | | | |
| Bread | UK | 157 | ND-2 | ND-4.5 |
| | Denmark | 33 | 4 | 0.8-12 |
| Kimchi | South Korea | 20 | 4 | ND-16 |
| Yoghurt | UK | 4 | - | ND |
| | Various | 9 | 1 | ND-1.3 |
| | Denmark | 19 | 0.2 | ND-0.3 |
| Cheese | Various | 17 | - | ND |
| Soy sauce | Japan | 48 | ND-16 | Nd-84 |

(FAO/WHO, 2006a)

ND: Is value at or below the limit

Note that the results have been aggregated into product groups with the same country assignment when possible

(a) This mean concentration excludes the single highest value reported, 6,131µg/kg

(b) ND: not detected, i.e results ≤ LOD/LOQ of the method used.

Many edible plants contain cyanogenic glucosides, with concentrations varying widely as a result of genetic and environmental factors (Ermans *et al*, 1980a; FAO/WHO, 1993a). Concentrations of hydrocyanic acid in some food stuffs are shown in table 2.2 (modified from Simeonova and Fishbeir, 2004).

**Table 2.2: Hydrocyanic acid concentrations in food products (mg/kg) and beverages mg/l)**

| Type of product | Hydrocyanic acid concentrations (mg/kg or mg/l) |
|---|---|
| Cereal grains and their products | 0.001 – 0.45 |
| Soy protein products | 0.07 – 0.3 |
| Soy bean hulls | 1.24 |
| Apricot pits | 89 – 2, 170 |
| Fruit juice (cherry, apricot, prune) | 1.9 – 4.6 |
| Cassava | 300 – 2, 360 |
| Sorghum (immature) | 2,400 |
| Bamboo (immature shoot tip) | 7,700 |
| Lima beans | 2,000 – 3,300 |

(Ermans *et al.*, 1980b; FAO/WHO, 1993b).

## CHEMICAL AND PHYSICAL PROPERTIES

Urethane, the ethyl ester of carbamate acid, $(NH_2COOC_2H_5)$ is a colourless, almost odourless, columnar crystal or white granular powder (Hirschback *et al*, 1948). The similarity of carbamate acid esters to urea restricted in the classification of these compounds. Urethane is readily soluble in water alcohol, and liquids (Mirvish, 1968; Jones *et al*, 1977b). At $25^0C$, 1g of urethane dissolves in 0.5g of water, at room temperature urethane volatizes, and at $103^0C$ urethane sublimes (WHO, 1974c).

## OTHER PHYSICAL PROPERTIES

Molecular weight                89.09g/mol

  Boiling point                $182^0C$

  Melting point               $48^0C$

  Density/specific gravity       0.9862 (water = 1)

  Vapour density             3.07 (air = 1)

Log and octanol/water partition coefficient (log $K_{ow}$) is - 0.15

  Vapour pressure            0.36 at $25^0C$

  Conversion factor          1ppm = 3.6mg

  (WHO, 1974d)

## 2.3 Anaesthesia

Urethane has been used for many years to produce trypnosis and narcosis in mammals, fish and amphibians (Rossoff, 1974; Green, 1982b). Urethane is specifically indicated for producers in which the investigator requires anaesthesia that lasts for hours with minor alterations in the patients physiologic parameters. At anaesthetic dosages (1 – 1.2g/kg body weight for rats) urethane has a wide margin of safety and causes minimal changes in blood pressure, aortic blood flow and blood-gas values (Folle and Levesque, 1976; Buelke-Sam *et al.*, 1978; De Wildt *et al.*, 1983).

When used in combination with alpha chloralose, urethane increases the solubility of chloralose, augments the activity of chlorase, and provides the necessary analgesia for surgical anaesthesia (Green, 1982c). Urethane will also suppress excess muscle activity, exaggerated spinal reflex activity and central nervous system stimulation commonly associated with chlorase anaesthesia (Hughes *et al.*, 1982a).

## 2.4 Mutagenic and Anti-Mitotic Evidence

When urethane is administered to *Drosophila*, mutagenic properties similar to those created with X-ray radiation have been observed (Auerbach, 1967a). Chromosomal breakage and recombination will occur and gene mutations develop without evidence of chromosomal breakage (Auerbach, 1967b).

At a dosage of 1g/kg IP, (which is the normal anaesthetic dose in most species), urethane will arrest cell division in the crypt of liberkuhn cells in mice (Bastrupt-madien, 1949a). Urethane is known to be cytotoxic to dividing cells, thereby decreasing mitotic activity. This finding may partially explain urethanes antineoplatic properties.

## 2.5 Absorption and Distribution

Following IP injection of radioactively labeled urethane into mice ($C^{14}$), no specific tissue localization occurs, urethane distributes to all tissues in the body (Bryan *et at*, 1957a). *In vivo*, urethane is metabolized to ammonia, carbon (iv) oxide and ethyl alcohol (Bryan *et al.,* 1957b; Boyland and Nery,1965). Approximately 90% of the radioactivity is detected in the expired carbon (iv) oxide and 5 – 10% is detected in urine (Bryan *et al*, 1957c). In mice, when urethane is applied directly to the skin, a sufficient amount is absorbed to induce a transient narcosis.

## 2.6 Carcinogenic and Immuno Suppressive Data In Humans

There are presently no case reports or epidemiological studies available on the carcinogenic potential of urethane in humans (WHO, 1974e). This may be related to the fact that urethane was used mainly as a chemotherapeutic agent in terminal cancer patients, long term follow-up data may not be available from such patients.

However, urethane is a known animal carcinogen and shows consequently be treated as a potential human carcinogen. Urethane is a well established immune suppressive agent in humans developing irreversible aplastic anaemia (Bother, 1949), Lymph node fibrosis, and fatal pneumonia secondary to agranulocytosis (Letterer, 1949) following treatment with urethane. When the immune suppressive findings are coupled with the knowledge that urethane is well absorbed across the skin, casual skin contact could obviously pose a serious health risk to humans.

## 2.7 Hazards

Ethyl carbamate is not acutely toxic to humans as shown by it's use as a medicine. Acute toxicity studies show that the lowest fatal dose in rats, mice, and rabbits equals 1.2grams/kg or more. When ethyl carbamate was used medicinally, about 50 percent of the patients exhibited nausea and vomiting and long time use led to gastroenteric hemorrhages. The compound has almost no odour and a cooling saline bitter taste.

Studies with rats, mice and hamsters has shown that ethyl carbamate will cause cancer when it is administered orally, injected, or applied to the skin but no adequate studies of cancer in humans caused by ethyl carbamate has been reported due to the ethical considerations of such studies.

However in 2007, the International Agency for research on Cancer raised ethyl carbamate to a group 2A carcinogenic, that is "probably carcinogenic to human",. One level below fully carcinogenic to humans. IARC has stated that ethyl carbamate can be reasonably anticipated to be a human carcinogen based on sufficient evidence of carcinogenicity in experimental animals". In 2006, the liquor control Board of Ontario in Canada rejected imported cases of sherry due to excessive level of ethyl carbamate. Alcoholic beverages, particularly certain stone-fruit spirits and whiskies, tend to contain much bigger concentrations of urethane. Heating (e.g, cooking) the beverage increases the ethyl carbamate content and some concern exists over shipping wines to overseas markets in containers that tend to overload. In addition, urethane has tendency to accumulate in the human body from a number of daily diet foods e.gs. alcohols, bread and other fermented grain products, soy sauce, orange juice and other foods. Hence, exposure risk to human health is increasingly evaluated on total ethyl carbamate content from the daily diet (WHO refers to this as "margin or exposure" or MOE).

## 2.8 Health Hazard Information

### Acute effects

➢ Acute exposure of humans to high level of ethyl carbamate may result in injury to the kidneys and liver and induce vomiting, coma or hemorrhages.

➢ Acute exposure of animals has been reported to cause bone marrow and central nervous system depression.

### Chronic effects

➢ No information is available on the chronic effects of ethyl carbamate in humans or animals.

➢ EPA has not established a reference concentration (RFC) or a reference dose (RED) for ethyl carbamate

### Reproductive/developmental effects

➢ Animal studies have reported effects on the developing fetus from maternal exposure to ethyl carbamate, including fetal abnormalities and fetal mortality.

➢ Living tumors have been reported in the offspring of mice injected with ethyl carbamate while pregnant.

## 2.9 Legislation of Ethyl Carbamate And Hydrocyanic Acid

### Ethyl Carbamate

There are currently no harmonized maximum levels for ethyl carbamate ion the European Union (EU). However, some member states and third countries recommend maximum levels for ethyl carbamate in alcoholic beverages (table 2.2)

**Table 2.3: Maximum levels for ethyl carbamate in alcoholic beverages**

| Country | Ethyl carbamate concentration µg/L in legislation | | | | |
|---|---|---|---|---|---|
| | Wine | Fortified wine | Distilled spirits | Sake | Fruit brandy |
| Canada | 30 | 100 | 150 | 200 | 400 |
| USA | 15 | 60 | | | |
| Czech Republic | 30 | 100 a) | 150 | 200 | 400 b) |
| France | - | - | 150 | - | 1000 |
| Germany | - | - | - | - | 800 |

a) Fruity wines and liqueurs

b) Fruity distilled and fruity, mixed and other spirits

(FAO/WHO, 2006b)

## 2.10 Current Results

Because of the sparse number of results reported to the JECFA for ethyl carbamate in some important beverage categories, EFSA in late September 2006 issued a call for submission of data on levels of ethyl carbamate and hydrocyanic acid in food and beverages. Seven EU member states (see table below) responded to EFSA's call for data on ethyl carbamate and submitted results covering analysis for 1998 to 2006. Three member states (Germany, France and Austria) also submitted results from cyanide testing. The sensitivity of the methods used for ethyl carbamate as reported by the LOD ranged from 0.1µg/kg to 400µg/kg (in few instances). The LOD for hydrocyanic acid varied from 10 to 100 µg/kg. During the evaluation of data, incomplete and duplicate records were omitted, as well as samples with a LOD for ethyl carbamate at risk assessment. The number of valid results incorporated in the further analysis is illustrated below. Of the 4,203 results for ethyl carbamate, 137 covered food samples and 4,066 covered alcoholic beverage samples. Testing for hydrocyanic acid included 715 results covering spirits other than fruit brandy and fruit brandy, and one further result covering wine. Ethyl carbamate was detected in 59% of food samples and 88% of alcoholic beverages samples. No food sample was tested for hydrocyanic acid. Test results for whisky from 1998 and 1999 were incorporated in the materials since very few whisky were received from the call for data covering 2000 to 2006.

**Table 2.4: The number of samples analysed for ethyl carbamate by respective member state for the years 1998 to 2006.**

| Country | 1998 | 1999 | 2000 | 2001 | 2002 | 2003 | 2004 | 2005 | 2006 | Total |
|---|---|---|---|---|---|---|---|---|---|---|
| Austria | 139 | 16 | 20 | 5 | 2 | 4 | 2 | - | - | **188** |
| Belgium | - | - | - | - | - | - | - | - | 11 | **11** |
| Czech Republic | - | - | - | - | - | - | - | - | 92 | **92** |
| France | - | - | 48 | 42 | 41 | 27 | 65 | 19 | 40 | **282** |
| Germany | 22 | - | 181 | 387 | 278 | 377 | 499 | 620 | 845 | **3,293** |
| Netherlands | - | - | - | - | - | - | - | 67 | - | **67** |
| United Kingdom | - | 205 | - | - | - | - | - | 149 | - | **135** |
| **Total** | **161** | **221** | **249** | **434** | **321** | **408** | **566** | **855** | **988** | **4,203** |

(FAO/WHO,2006c)

## 2.11  FACTORS INFLUENCING THE CONCENTRATION OF ETHYL CARBAMATE

Since the identification of high levels of ethyl carbamate in alcoholic beverages there has been a considerable focus on reducing levels by a range of different methods (Lachenmier *et al.*, 2005). The key to successful prevention and control has been the identification of the main precursor substances responsible for the formation of ethyl carbamate in food and beverages, together with an understanding of the influence of the main external factors of light (for spirits only), time and temperature.

Over the past years, major reductions in concentrations of ethyl carbamate have been achieved by reducing the concentration of the main precursor substances in the food or beverage and by reducing the tendency of these substance to react to form cyanate, e.g. by the exclusion of light from bottled stone- fruit brandies. The reaction mechanisms that give rise to ethyl carbamate vary significantly between foods (Ough *et al.*, 1988). In the case of wine, the use of urea as a yeast nutrient can result in an elevated level of ethyl carbamate in the finished product while light has little influence. Ethyl carbamate formation increases overtime with the reaction rate being exponentially accelerated at higher or elevated temperatures. Urea is formed then the wine yeast metabolizes arginine, a major alpha-amino acid in grape juice available to yeast. This reaction is yeast strain dependent. Yeasts differ in their ability to produce urea and to reuse urea secreted into the must wine. Lactic and bacteria also metabolize arginine and liberate citrulline, an amino acid, which then reacts with ethanol to form ethyl carbamate. Over fertilized vineyards, in general, will yield wines with higher urea and thus ethyl carbamate concentrations (Butzke and Bisson, 1997), much higher ethyl carbamate concentrations were detected in spirits derived from stone fruit like cherries plums, Mirabelles, or apricots (Battaglia et al., 1990b; Zimmerli and Schlatter, 1991b) the formation of cyanogenic glycosides such as amygdalin in stone fruit by enzymatic action leads to the generation of cyanide, which is the most precursor of ethyl carbamate in these spirits. The wide range of ethyl carbamate concentrations in stone fruit spirits reflects its light induced and time dependent formation after distillation and during storage (Audrey, 1987; Mildau et al., 1987; Baumann and Zimmerli; 1988; Zimmerli and Shlatter, 1991c; Suzuki et al., 2001).

This is a major problem in that ethyl carbamate continues to be formed in spirits during storage so analytical results might not provide an accurate estimate of the potentials level which might be encountered at a later point in time. Some methods thus incorporate a light including step to determine the ethyl carbamate formation potential. Results reported to EFSA covered the use of methods both with and without light induction. However, results presented here only include "actual" levels detected without light induction.

## 2.12    Relationship    Between    Ethyl    Carbamate    And Hydrocyanicacid

Since hydrocyanic acid is a precursor to the formation of ethyl carbamate it has been hypothesised that are relationship exists between the respective amounts of ethyl carbamate and hydrocyanic acid found in alcoholic beverages to test this hypothesis, a linear regression between the log transformed amount of hydrocyanic acid and ethyl carbamate detected in 260 samples of alcoholic beverages with varies for both compounds above the respective limit of detection was undertaken.

A statistical analysis predicted the following linear regression equation also lustrated in figure below.

Log (ethyl carbamate µg/kg) = 1.877+ 0.284* log (hydrocyanic acid µg/kg )

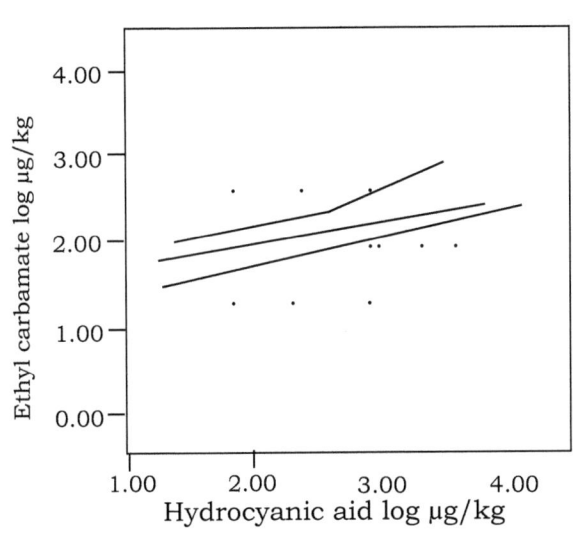

**Figure 2:** distribution of results in relating Hydrocyanic acid and ethyl carbamate log transformed concentrations and illustration of the linear regression equation.

The regression coefficient was 0.284 with a correlation coefficient (R) of 0.419 for the log transformed concentrations. There is thus some relationship between the two compounds judged by the correlation coefficient, but with some considerable spread in the date.

A coefficient of determination ($R^2$), indicating the ratio of explained variation to the total variation of 0.175 showed that only 17.5% of changes in ethyl carbamate levels can be explained by changes in hydrocyanic acid concentrations.

This is not an expected finding since ethyl carbamate formation is multi-factorial, among many other factors time and light dependent in most case, and there was a lack of detailed knowledge about the handling the product after Manufacturing .

However, comparing the distribution of samples with ethyl carbamate concentrations above and below 200 or 400 µg/kg for different levels of hydrocyanic acid provided a better fit.

**TABLE 2.5:** Distribution of samples across given limits for ethyl carbamate of 200 or 400 ng/kg for different levels of hydrocyanic acid.

| Hydrocyanic acid ng/kg | Ethyl carbonate ng/kg beverages | | | | |
|---|---|---|---|---|---|
| | **ALL** | **<200** | **≥200** | **<400** | **≥ 400** |
| < 5,000 | 645 | 278(43%) | 367(57%) | 405(63%) | 240(37%) |
| ≥ 5,000 | 71 | 4(6%) | 67(94%) | 14(20%) | 57(80%) |
| ≥ 10,000 | 42 | 2(5%) | 40(95%) | 5(12%) | 37(88%) |
| ≥ 20,000 | 12 | 0 | 12(100%) | 1(8%) | 11(92%) |
| ≥ 28,000 | 5 | 0 | 5(100%) | 1(20%) | 4(80%) |
| ≥ 40,000 | 2 | 0 | 2(100%) | 0 | 2(100%) |

(FAO/WHO, 2006d)

## 2.13  Sources of Ethanol Under Study Ethanol From Cassava

The production of ethanol from sugar, cellulose, and converted has been practiced for a long time and up till date. In Nigeria, local production of ethanol from cereals like maize, guinea corn, millet, other starchy substrates, and cellulose is as old as the country itself. Apart from food and pharmaceutical uses, ethanol is finding itself alternative uses for biofuel in most of the developed world for the following reason:

- It is an excellent raw material for synthetic chemicals.
- It is not poisonous
- It does not cause air pollution and environmental hazard.

# CASSAVA ETHANOL PRODUCTION PROCESS FLOWCHART

Cassava flour
Water and alpha-amylase enzyme
⇩

Liquefaction
(90-95$^0$C,PH 4-4.5)400rpm
⇩

Saccharification
55-65$^0$C,pH4-4.5
Glucosidase enzymes
⇩

Coaling
(30-33$^0$C)
⇩

Fermenter
(yeast added, $CO_2$ out)
⇩

Distillation
(feed recovery)
⇩

Ethanol

All alcoholic beverages, including those produced by carbonic maceration, are produced by ethanol fermentation by yeast. Wine and brandy are produced by fermentation of the natural sugars present in fruits, especially grapes. Beer and whisky are produced by fermentation of grain starches that have been converted to sugar by the enzyme amylase, which is present in grain kernels that have been germinated. Amylase treated grain or amylase- treated potatoes are fermented for the production of vodka. Rum is produced by fermentation of cane sugar. In all cases, the fermentation must take place in a vessel that allows carbon (iv) oxide to escape, but prevents outside air from coming in, as exposure to oxygen would prevent the formation of ethanol.

In the United States the main feedstock for the production of ethanol is currently corn. Approximately 2.8gallons of ethanol are produced from one bushel of corn (0.42 liter per kilogram). While much of the corn turns into ethanol, some of the corn also yields by products such as DD GS (distillers dried grains with soluble) that can be used as feed for livestock and thus starches are cheapest and has the highest energy content.

## 2.14 Ethanol from Fuel Petroleum

Ethanol fuel is ethanol (ethyl alcohol), the same type of alcohol found in alcoholic beverages. It is most often used as a motor fuel mainly as a biofuel addictive for gasoline.

Ethanol fuel is widely used in Brazil and in the United States, and together both counties were irresponsible for 88% (percent) of the world's ethanol fuel production in 2010. Bio ethanol is a form of renewable energy that can be produced from agricultural feedstocks. It can be made from very common crops such as sugar cane, potato, manioc and corn. However there has been considerable debate about how useful bioethanol will be in replacing gasoline. Concerns about it's production and use relate to increased food prices due to the large amount of arable land required for crops as well as the energy and pollution balance of the whole cycle of ethanol production, especially from corn. Recent developments with cellulosic ethanol production and commercialization may allay some of these concerns.

Cellulose ethanol offers promise because cellulose fibres, a major and universal component in plant cells walls, can be used to produce ethanol. According to the international energy agency, cellulose ethanol could allow ethanol fuels to play a much bigger role in the future than previously thought.

Ethanol being a renewable energy source because the energy is generated by using a resource, sunlight, which cannot be depleted. Creation of ethanol starts with photosynthesis causing a feedstock, such as sugarcane or corn to grow. These feedstocks are processed into ethanol. About 5% of the ethanol produced in the world in 2003 was actually a petroleum product. It is  made by catalytic hydration of ethylene was sulfuric acid as the catalyst  it can also be obtained via ethylene or acetylene from calcium carbide, coal oil gas and other sources. Two million, tons of petroleum derived ethanol are produced annually. The principal suppliers are plants in the United States, Europe, and South Africa. Petroleum derived ethanol (synthetic ethanol) is chemically identical to bio ethanol can be differentiated only by radio carbon dating.

Bio-ethanol is usually obtained from the conversion of carbon based feedstock Agricultural feedstock's are considered renewable because they get energy from the sun using photosynthesis, provided that all mineral required for growth such as nitrogen and phosphorus are returned to the land. Ethanol can be produced from a variety of feedstock's such as sugarcane, bagasse, miscanthus, sugar beet, sorghum, grain sorghum, switch grass, barley, hemp, kenaf, potatoes, sweet potatoes, cassava, sunflower, fruit, molasses, corn, Stover, grain, wheat, straw, Cotton, other biomass as well as many types of cellulose waste and harvestings, which ever has the best well–to–wheel assesement.

Currently, the first generation processes for the production of ethanol from corn use only a small part of the corn plant: the corn kernels are taken from the complaint and only the starch, which represents about 50% of the dry kernel mass, is transformed into ethanol.

## PRODUCTION PROCESS

### Distillation

For the ethanol to be usable as a fuel, water must be removed. Most of the water is removed by distillation, but the purity is limited to 95-96 % due to the formation of a low–boiling water–ethanol azoetrope. The 95.6% m/m (96.5%v/v) ethanol, 4.4 %m/m (3.5%v/v) water mixture may be used as fuel alone but unlike anhydrous ethanol, is immiscible in gasoline, so that the water fraction is typically removed in further treatment in order to burn in combination with gasoline in gasoline engines.

### Dehydration

There are basically five dehydration processes to remove the water from an azeotropic ethanol/water mixture. When these components are added to the mixture, it forms a heterogeneous azeotropic mixture in vapour – liquid-liquid equilibrium, which when distilled produces anhydrous ethanol in the column bottom, and a vapour mixture of water and cyclohexane/benzene. When condensed this becomes a two-phase liquid mixture. Another early method, called extractive distillation, consists of adding a ternary component which will increase ethanol's relative volatility. When ternary mixture is distilled it will produce anhydrous ethanol on the top stream of the column.

## 2.15 Ethanol Production Process from Sugarcane

The first step of the ethanol production processes takes place in the fields where sugarcane is harvested. The sugarcane plant is composed by roofs, stalk, tops and leaves. The stalks contain most of the sugars, thus being, the fraction of intrest in industrial processing. Upon mechanical harvest, most of the leaves and tops, which constitute the so called "sugar cane trash", are separated from the stalks and left on the fields, increasing soil protection and incubating the growth of weed and other plant species sugar cane received in the factory, comprised mainly by it's stalks, contain water, fibre, sugars, impurities and dirt (sand, simulated as silica).

All reducing sugars were simulated as dextrose; fibre (12% of the sugar cane) is comprised by cellulose, hemicellulose and lignin (Wooley and Putsche, 1996), "Impurities were simulated as aconites acid and potassium salt, since those account for the majority of the impurities (Mantelatto, 2005); Minerals were simulated as $K_2O$ and dirt as $S_iO_2$. Dirt is dragged along with the sugar cane during harvest; the other components are part of sugar cane structure.

In this work an autonomous distillery, that is a factory in which sucrose is entirely used as raw material for ethanol production, is considered for the production of $1000m^3$ per day of drops ethanol, using sugar cane juice as raw material. An input of 493t of sugarcane per hour is necessary. This is compactable with large scale units of operating nowadays.

Ethanol production from sugar cane is comprised by the following steps: cleaning of sugar cane and extraction of sugars, huge treatment, concentration and sterilization; fermentation, distillation, dehydration. A detailed description of the simulation of each of these steps is presented and the block flow diagram is depicted. Cleaning of sugarcane, extraction of sugars and juice treatment in order to remove part (70%) of the dirt dragged along with the sugarcane from the fields, a dry- cleaning system is used. Sugar extraction is carried out using mills, in which water at the rate of 28% of sugarcane flow is used to enhance sugar recovery in a process termed imbibitions (Chen and Chou, 1993a). Sugars recovery in the mills is considered equal to 96%. In the mills, sugars cane juice and bagasse are obtained. Sugar cane juice which contains the sugars is fed to the juice treatment operations, while sugarcane bagasse (50% humidity) is burnt in boilers for generation of steam and electrical energy. Sugar cane juice contains impurities, such as minerals, salts, acid, dirt and fibre, besides water and sugars in order to be efficiently used as a raw material for ethanol production through fermentation, those impurities must be removed; thus the juice is submitted to physical and chemical treatments.

Screens and hydrocylones are used in the physical treatment, where the majority of fibre and dirt particles are removed (Chen and Chou, 1993b). In subsequent chemical treatment, phosphoric acid is added to sugar cane juice, to increase juice phosphates content and enhance impurities removal during settlement, followed by the first heating operation in which juice temperature increases from 30 to 70$^{\circ}$C. Besides impurities, the mud obtained in the settler contains sugars, thus a filter is used to enhance sugars recovery.

**Juice Concentration and Sterilization**

Clarified juice contains around 15wt% diluted solids, so it must be concentrated before fermentation in order to achieve an adequate ethanol content that allows reduction of energy consumption during product purification steps.

Juice is sterilized prior feeding the fermentation reactor in order to avoid contamination, which would decrease fermentation yields. During sterilization, juice is heated up to 130$^{\circ}$C during about 30minutes and then rapidly cooled down to fermentation temperature.

**Fermentation**

Yeast supervision containing about 28% yeast cells (v) is fed to the fermentation reactor, along with sterilized juice. Yeast accounts for approximately a quarter of the reactor capacity. During the fermentation reactions, sucrose is hydrolyzed into fructose and glucose, which are converted into ethanol and carbondioxide, as represented in equation 1

$$C_6H_{i2}O_6 \longrightarrow 2C_2H_6O + 2CO_2 \qquad (1)$$

A conversion reactor was used in the simulation, and the conversion for equation (1) is equal to 90.48%, based on sugar consumption. Some by products are formed as well, as a result of parallel fermentation reactions, cell growth and impurities in the sugarcane juice, among other factors.

The fermentation temperature represents ascetical step in ethanol production, since higher temperature affect yeast behaviour, diminishing the ethanol content of the final wine, which increase the consumption of energy during distillation as well as the volume of vinasse per volume of ethanol produced (Dias *et al*, 2007) fermentation was considered to be performed at 28°c, which is considerably lower than temperatures usually used in industry (around 34°C).

## Distillation and Dehydration

Hydrous ethanol is usually produced through convectional distillation, whose configuration containing between 92.8 and 93.5wt% ethanol, is usually achieved in Brazilian industries in a series of distillation and rectification columns operating under atmospheric pressure production of anhydrous ethanol (99.5% ethanol, mass basis) is carried out using an extractive distillation process with mono ethylene glycol (MEG). The extractive distillation process with MEG was chosen because of the low energy consumption on reboilers and production of high quality bio ethanol (Meirelles *et al*, 1992). In this process two distillation columns are used: extractive and recovery. The extractive column operates under atmospheric pressure, while the recovery column operates under low pressure (20kpa) in order to avoid solvent decomposition and high temperatures. Hydrous ethanol vapors are fed near the bottom of the extractive column, while solvent (MEG) is fed near it's top. Anhydrous ethanol is obtained as the top product, while a recovery column is used to produce pure solvent, which is cooled and recycled to the extractive column. Temperature on top of the extractive column is approximately 78°c, so it is possible to integrate the extractive column condenser to column reboiler, producing the remaining heat necessary to the adequate process operation.

## SUGARCANE ETHANOL PRODUCTION PROCESS FLOW CHART

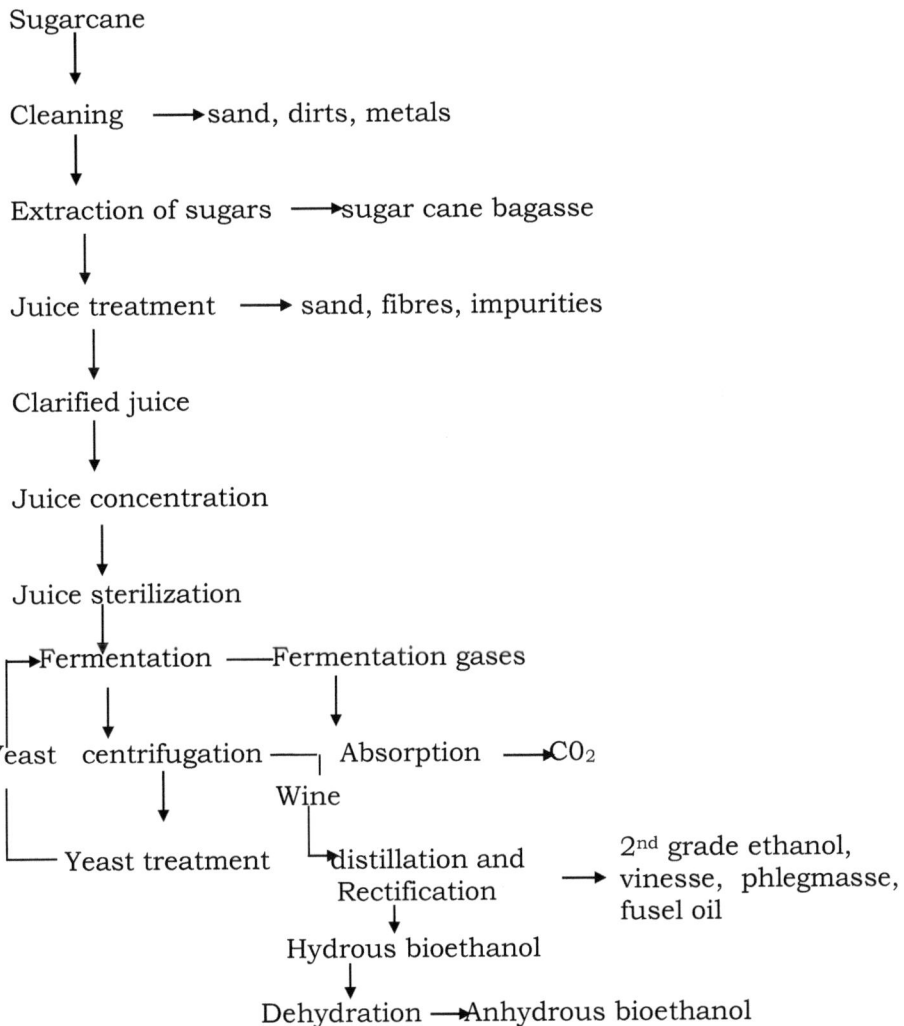

Sugarcane

↓

Cleaning ──→ sand, dirts, metals

↓

Extraction of sugars ──→ sugar cane bagasse

↓

Juice treatment ──→ sand, fibres, impurities

↓

Clarified juice

↓

Juice concentration

↓

Juice sterilization

↓

Fermentation ──── Fermentation gases

↓

Yeast  centrifugation ── Absorption ──→ $CO_2$

               ↓      Wine

Yeast treatment  distillation and ──→ 2nd grade ethanol, vinesse, phlegmasse, fusel oil

Rectification

↓

Hydrous bioethanol

↓

Dehydration ──→ Anhydrous bioethanol

# CHAPTER THREE

## 3.0 MATERIALS AND METHODS

### 3.1 Source of Materials

The three spirit samples from different materials of production; cassava, sugarcane and palm wine represented as C,S,P were purchased from major distributors in Aba, Abia State Nigeria.

All equipments used were obtained from the departments of Food Science and Technology and Industrial Chemistry of the Federal University of Technology Owerri, Imo State.

Other materials and reagents were of analytical grade and supplied by the same departments.

### CHEMICALS/REAGENTS

Orthophosphoric acid (concentrated)

Sodium hydroxide pellets

Silver nitrate, 0.01N standard volumetric solution

Potassium iodide (kI) solution (50 g/l)

Ammonia solution    (6N)

Distilled water

All the above reagents were of BDH (British drug house) type.

## DETERMINATION OF HYDROCYANIC ACID CONTENT

Hydrocyanic acid is amongst one of the precursors of ethyl carbamate thus hydrocyanic acid expressed in milligram of HCN of sample is, equal to

$$0.54 \ (V_0\text{-}V1) \ \times \frac{250}{120} \times \frac{100}{m} = 112.5 \frac{(V_0\text{-}V_1)}{m}$$

Where m = mass of test portion (20)g

$V_0$ = volume in ml of 0.01N silver nitrate solution used for the sample determination

$Vi$ = volume in ml of 0.01N silver nitrate solution used for the blank test.

## METHOD OF DISTILLATION/DISTILLATION PROCESS

This method is commonly used in separating solvents from solutions especially when the solvent is needed. It can be used to separate the more volatile from less volatile substances. Water (solvent) can be obtained from sodium chloride solution by distillation.

Distillation can also be employed in separating mixture of liquids of different boiling points. This technique of separation depends on the fact that pure substances have definite boiling points at a given pressure. This method involves heating the mixture in a flask to vapourise the solvent. The vapour passes along a condenser called a liebig condenser which converts vapour to liquid while distillating. The condensed steam is called the distillate or condensate which is also collected in a receiver. The solute and other impurities are left behind in a distillation flask.

Distillation is used in gin distilleries and water distilleries for the manufacture of gin and distilled water respectively.

In this work distillation was done so as to get or to determine the hidden urethane in the ethyl alcohol from different samples since ethyl carbamate cannot be directly detected unless by using GC/MS (Gas chromatography/mass- spectrometry) and thus which is too expensive and scarce.

## Determination of Ethyl Carbamate (Urethane)

Ethyl carbamate content of chemical sample (ethyl alcohol) were determined first by determining the glycosidic cyanide content of the ethanol as described by FAO (1986) with little modification.

A 20ml of ethyl alcohol sample was transferred into 1 litre distillation flask then 200ml of distilled water and 10ml of orthophosphoric acid added. The flask was stopped, mixed and left for 12 hours (overnight) in an incubator (Gallykamp) at 38$^0$C.

The flask was fitted to the distillation apparatus and distilled into 20ml and was transferred into 250 ml volumetric flask and diluted to the mark using distilled water.

A 100ml of the aliquot was pipetted into a beaker and 2ml of 5% potassium iodide solution and 1ml of 6N ammonia solution added. The mixture was titrated with 0.01N silver nitrate solution until permanent turbidity was observed. For easy recognition of the end point of titration, it is recommended that a black background should be used. A blank test under the same condition as in the sample determination but replacing the distillate with distilled was carried out.

The table of result for cyanide and corresponding Ethyl carbamate concentration by (Aylott *et al.*,1990b) was used to produce a regression equation as shown (Cass, 1971).

**Table 3.1: Cyanide and corresponding ethyl carbamate concentration**

Measurable cyanide Corresponding ethyl carbamate (mg/120g) (mg/100g)

| X | Y | XY | X² | Y² |
|---|---|---|---|---|
| 0.5 | 1.00 | 0.50 | 0.25 | 1.00 |
| 3.75 | 0.75 | 2.81 | 7.90 | 0.56 |
| 4.50 | 1.25 | 5.63 | 31.70 | 1.56 |
| 5.50 | 1.75 | 9.63 | 92.74 | 3.06 |
| 13.00 | 2.25 | 29.25 | 855.56 | 5.06 |
| $\Sigma x = 27.25$ | $\Sigma y = 7.00$ | $\Sigma xy = 47.82$ | $\Sigma x2 = 988.15$ | $\Sigma y2 = 11.24$ |

From ($y = a + bx$)

Skipping some mathematical steps; the required values of a and b may be found by solving pair of simultaneous equations.

$$\Sigma y = na + b \Sigma x$$

$$\Sigma xy = a\Sigma x + b\Sigma x^2$$

From table 3:1, n = 5, therefore b = 0.01, and a = 1.346.

The required regression equation is therefore

Y = a + bx

i.e y = 1.346 + 0.01x

Therefore, if the value of x (i.e. measurable cyanide) is known, the corresponding value of y (i.e ethyl carbamate) can be calculated by substituting the value of x in the regression equation above.

# CHAPTER FOUR

## 4.0 RESULTS AND DISCUSSION

All the spirit samples tested contained measurable cyanide ranging between 16.59 – 36.84 mg/120ml of the sample and their respective ethyl carbamate content 1.15 – 1.71mg/120ml of the samples (table 4.3).

The glycosidic cyanide of all the samples analysed were within safe level used in beer production and then allowed in human foods (5mg/100g), and within such concentration of cyanide in food items, the foods or drinks are free from urethane (Aylott *et al*, 1990c).

However the beverages (i.e beers and spirits) should not be stored under adverse conditions such as in the sun as most wholesalers, retailers, distributors do, and certainly storage period should not be prolonged to avoid possible conversion of cyanide to urethane which is possible with subsequent reaction overtime in the presence of heat (Wardlaw and Insel, 1996b).

**Table 4.1: pH of the ethyl alcohols samples from different sources**

| SAMPLE | SOURCE | pH |
|--------|-----------|-----|
| C | Cassava | 3.8 |
| P | Palm wine | 4.0 |
| S | Sugarcane | 3.5 |

**Table 4.2: The alcohol content of the various ethanol (ethyl alcohol) by volume**

| SAMPLE | ALCOHOL CONTENT (%V/V |
|--------|------------------------|
| C | 96.0% |
| P | 96.6% |
| S | 96.0% |

**Table 4.3: Measureable cyanide content and the corresponding ethyl carbamate (urethane) content of the ethyl alcohol samples**

| Ethyl alcohol sample | Measureable cyanide content (mg/120ml) | Ethyl carbamate (mg/120ml) |
|---|---|---|
| $C_1$ | 20.53 | 1.55 |
| $C_2$ | 23.23 | 1.57 |
| $P_1$ | 16.59 | 1.51 |
| $P_2$ | 19.18 | 1.53 |
| $S_1$ | 33.75 | 1.68 |
| $S_2$ | 36.84 | 1.71 |

The required regression equation is therefore $y = a + bx$

$$1.346 + 0.01x$$

Therefore if the value of x (i.e measurable cyanide) is known, the corresponding value of y (i.e ethyl carbamate) can be calculated by substituting the value of x in the regression equation above.

Table 4.3 exhibits the cyanide against corresponding urethane content of the ethyl alcohol samples.

# CHAPTER FIVE

## 5.1 CONCLUSION

Nigeria alcoholic beverage especially the spirits and beers are not completely free from urethane. However, the traces of urethane observed are at safe levels. I advise that more research interest should be paid on this area of research since this work have provided the impetus and framework.

## 5.2 RECOMMENDATION

In this project work, ethyl carbamate was not measured directly and as such, it took a very long period of time to redistill the samples titrate and to get the actual result.

I recommend that the health authority (NAFDAC) be put on its toes regarding the presence of urethane in all the fermented beverages since this work has proved that they are found at different levels.

I also suggest mitigation measures should be taken to reduce the levels of ethyl carbamate in certain alcoholic beverages such as fruit brandies such measures should include focus on hydrocyanic acid and other precursors of ethyl carbamate to prevent the formation of ethyl carbamate during shelf life of those products.

Wineries should educate and encourage the shipper, distributors, wholesaler, and retailer to minimize heat exposure by use of appropriate insulated containers, shipping schedules and storage facilities.

# REFRENCES

Andrey, D. (1987). A simple gas chromatography method for the

determination of ethyl carbamate in spirits. Zeltschrift fuer Leben smittel. Untersu Chung Und-Forschung. 185: 21-23.

Anon, (1974). Monograph on the evaluation of the carcinogenic risk

of chemicals to man. International Agency for Research on Cancer (I.A.R.C), Lyon. 7: 11-15.

Auerbach, C. (1967). The chemical production of mutations.

Science. 158: 1141-1147.

Aylott, R.I., Cochrane, G.C., Leonard, M.J. MacDonald, L.S.,

Mackenzie, W.M. and McNeish A.S. (1990). Ethyl carbamate formation in grain-base whisky. Part 1: Post distillation ethyl carbamate formation in maturing grain whisky. Journal of institute of Brewing. 96: 213-221.

Aylott, R.I., McNeish, A.S and Walker, D.A. (1987). Ethyl carbamate

formation in grain-based spirits. Journal of institute of Brewing. 93: 382-388.

Bastrup-Maiden, P.C. (1949). Action of mitotic poisons in vitro.

Effect of urethane on division of fibroblasts. Acta pathologia et microbiological Scandinavia. 26: 93-112.

Battaglia, R., Conacher, H.B.S and Page., B.D. (1990). Ethyl

carbamate (urethane) in alcoholic beverages and foods – a review. Food additives and contaminants. 7: 477-496.

Baumann, U. and Zimmerli, B. (1988). Accelerated formation of

ethyl carbamate in spirits. Milleilung Gebiete Lebensmittel Hygenik. 22: 175-185.

Benson, R.W and Beland, F.A. (1997). Modulation of urethane

(ethyl carbamate) carcinogenicity by ethyl alcoholic: a review. International journal of toxicology. 16: 521-544.

Bryan, C.E, Skipper, H.E and White, L. (1957). Carbamate in

chemotherapy of Leucemia IV. The distribution of radioactivity in tissue in mice following injection of carbonyl-labelled urethane. Journal of biological chemistry. 177: 941-950.

Buelke-Sam. J., Holson, J.F., Bazare, J.J. and Young, J.F. (1978).

Comparative stability of physiologic parameters during sustained anaesthesia in rats: Labouratory Animal Science. 28: 157-162.

Butzke, C.E. and Bisson, L.F. (1997). Ethyl carbamate preventative

action manual. US Food and Drug Administration, Washington D.C, USA. Available at URL:http://Vm.cfsan.fda.gov/^frf/ecaction.html

Cass, T. (1971). Statistical method in management. Camelot press

Ltd, London and Southampton. Pp. 120-124.

Chen, J.C.P and Chou, C.C., (1993). Cane sugar hand book: a

    manual for cane sugar manufactures and their chemists. (John Wiley and sons), Pp. 1090.

Conacher, H.B.S and page, B.D. (1986). Ethyl carbamate in

    alcoholic beverages: A Canadian case history. Proceedings of Euro food Tox 11, European Society of toxicology, Schwerzenbach, Switzerland. Pp237-242.

Cook, R., McCaig, N., Mcmillan, J. and Lumsden, W. (1990). Ethyl

    carbamate formation in grain-based spirits part III: The primary source. Journal of institute of Brewing. 96: 233-244.

Daennis, M.J., Howarth, N., Key, P.E., Pointer, M., and Massey,

    R.C. (1989). Investigation of ethyl carbamate levels in some fermented foods and alcoholic beverages. Food addictives and contaminants. 6: 383-389.

Dewildt, D.J., Hillen, F.C., Rauws, A.G. and Sangster, B. (1983).

    Etomidate-anaesthesia, with and without fentanyl compared with urethane-anaesthesis in the rat. British journal of pharmacology. 79: 461-469.

Dias, M.O.S., Maciel Filho, R. and Rossell, C.E.V., (2007). Efficient

    cooling of fermentation vats in ethanol production - part 1. Sugar journals. 70: 11-17.

Dipalma, J.R. 91965). Pharmacology in medicine, 3rd.ed New York:

    McGraw Hill.

Eimans, A.M., Mbulamoko, N.M., Delange, F. and Ahluwalia, R.

(1980). Role of cassava in the etiology of endemic goiter and cretinism. International Development research Centre, 182pp, Ottawa, Ontario.

FAO/WHO (Food and Agriculture Organization of the United

Nations/World Health Organization). (2006). Safety evaluation of certain contaminants in food. Prepared by the sixty fourth meeting of the joint FAO/WHO Expert Committee on Food Addictives (JECFA) FAO Food Nutr. Pap. 82: 1-778.

FAO/WHO (Food and Agriculture Organization of the United

Nations/World Health Organization). (1993). Cyanogenic glycosides. In: toxicological evaluation of certain food additives and naturally occurring toxicants. 39th meeting of the joint FAO/WHO Expert Committee of food Additives (WHO Food Additives Series 30) World Health Organization, Geneva.

Finar, I.L. (1951). Organic chemistry. Vol. 1: Fundamental and

principle. Longman, Green and Co. Ltd., London. Pp. 383-384.

Folle, L.E. and Levesque, R.I. (1976). Circulatory, respiratory, and

acid base balance changes produced by anesthetics in the rat. Acta Biol med Germ. 45: 605-612.

Green, C.J. (1982). Animal Anaesthesia. Laboratory animal handbook 8. Colchester, London: Laboratory Animals Ltd.

Haddow, A and Sexton, W.A. (1946). Influence of carbamic esters

> (urethanes) on experimental animal tumors. Nature.
> 157: 500-503.

Harry, L., Riffkin, R., David, H and Steven, M. (1989). Ethyl

> carbamate formation in the production of whisky.
> Journal of institute of brewing. 95: 115-119.

Hirschback, J.S. Lindert, M.C, Chase, J. and Calvery, T. (1948).

> Effects of urethane in the treatment of leukemia and
> metastatic malignant tumors. Journal of National
> Cancer Institute. 5: 415-417.

Hoover, J.E. (1970). Remmgton's pharmaceutical sciences, 14[th]ed.

> Mac publishing Co.

Hughes, E.W., Martin-Body, R.L., Sarelius, I.H. and Sinclair, J.D.

> (1982). Effects of urethane-chloralose anesthesia on
> respiration in the rat. Clinical and experimental
> pharmacology and physiology. 9: 119-127.

IARC (international Agency for Research on Coancer) (2007).

> International Agency for research. Volume 96: Alcoholic
> beverage consumption and ethyl carbamate (urethane)
> 6-13 February 2007 1-5. World Health Organisation,
> Lyon,                                        France.
> http://monographs.iarc.fr/ENG/Meetings'/Vol96-
> summary.pdf.

Joness, L.M., Booth, N.H. and McDonald, L.E. (1977). Vertinary

> pharmacology and therapeutics 4th ed. Ames: Lowa
> Stata university press.

Kim, Y.K.L, Koh, E., Chung, H.J and Kwon, H. (2000).

Determination of ethyl carbamate in some fermented Korean foods and beverages. Food additives and contaminants. 17: 469-475.

Letter, E. (1949). Anatomicro pathologic observation of cadavers of

patients treated with urethane (abstract). 139: 125-126.

MacKenzie, W.M., Clyne, A.H. and MacDonald, L.S. (1990). Ethyl

carbamate formation in grain-based spirit. Part 11: identification and determination of cyanide related species involved in ethyl carbamate formation in Scotch grain whisky. Journal of institute of brewing. 96: 223-232.

Mantelatto, P.E, (2005). Study of the crystallization process of

impure sucrose solutions from sugarcane by cooling, Msc Dissertation (School of chemical Engineering, Federal University of Sao Carlos, Portuguese).

McGill, D.J, and Morley, A.S. (1990). Ethyl carbamate formation in

grain spirits part IV: Radiochemical studies. Journal of institute of Brewing. 96: 245-246.

Meirelles, A., Weiss, S. and Herfurth, H., (1992), Ethanol

dehydration by extractive distillation. Journal of chemical Technology and Biotechnology. 53: 181-188.

Mirish, S.S. (1968). The carcinogenic action and metabolism of urethane and n-hydroxy urethane. Advances in cancer research. 11: 1-42.

Mudau, G., Preub, A., Frank, W. and Heering, W. (1987). Ethyl

carbamate (urethane) in alcoholic beverages: improved analysis and light-dependent formation. Deutsche Lebensmittel-Rundschau. 83: 69-74.

Ough, C.S (1976). Ethyl carbamate formation in grain whisky.

Journal of Agricultural food chemistry. 24: 328-331.

Ough, C.S., Crowell, E.A. and Gutlove, B.R. (1988). Carbamyl

compound reactions with ethanol. AM. J. Enol. Vitic, 39:3, 239-242.

Paterson, E., Haddon, A., Thomas, I.A. and Watakinson, J.M.

(1946). Leukaemia treated with urethane compared to deep X-ray therapy. Lancet, 677-683.

Rossoff, I.S. (1974). Hand book of vertinary Drugs. New York:

Springer 8. Colchester, London: Laboratory Animals Ltd.

Schlather J. and Lutz, W.K. (1990). The carcinogenic potential of

ethyl carbamate (urethane): Risk assessment at human dietary exposure levels. food and chemical toxicology. 28:205-211.

Sen, N.P., Seaman, S.W. and Weber, D. (1992). A method for the

determination of methyl carbamate and ethyl carbamate in wines. Food additives and contaminants. 9: 149-160.

Sen, N.P., Seaman, S.W., Boyle, M. and Weber, D. (1993). Methyl

carbamate and ethyl carbamate in alcoholic beverages and other fermented foods. Food chemistry. 48: 359-366.

Simeonova, F.P. and Fishbein, L. (2004). Hydrogen Cyanide and

cyanides: human health aspects. Concise international chemical assessment document 61. World health Organization, Geneva.

Suzuki, K., Kamimura, H., Ibe, A., Tabata, S., Yasuda, K. and

Nishijima, M. (2001). Formation of ethyl carbamate in Umeshu (plu, liqueur). Shokuhin Elseigaku Zasshi. 42:354-358.

Templeton, W.G and Sexton, W.A. (1945)> effects of some aryl

carbamic esters and related compounds upon cereals and other plant species. Nature 156, 630.

Wardlaw, G.M. and Insel, P.M. (1996). Perspectives in nutrition.

3rded. McGraw Hill, New York. Pp. 728-730.

Wooley, R.J. and Putsche, V., (1996). Development of an ASPEN

PLUS physical property Database for biofuels components, Report No. NREL/MP-425-20685 (National Renewable Energy Laboratory, Golden, Colorado). Available online at: http://www.p2pays.org/ref/22/21210.pdf, retrieved on 12 march 2007.

World Health Organization (1974). International Agency for research on cancer monographs on the evaluation of carcinogenic risk of chemicals to man. 7: 111-140.

Zimmerli, B. and Schlatter, J. (1991). Ethyl carbamate: analytical

methodology, occurrence, formation, biological activity and risk assessment. Mutation research. 259: 325-350.

# APPENDIX 1

## EQUIPMENT AND APPARATUS

Analytical balance (Tara Shandon Sauter)

Cork opener

Measuring cylinder

I litre distillation flask

Lie big condenser

Volumetric flasks

Incubator maintained at $38^{0}C \pm 2^{0}C$

Electric heater

Transparent glass cups

Beakers

Conical flasks

Burettes

Pipettes

Steam distillation apparatus, G/G neck

Wooden corks

Cottons wool

Aluminum foil

**REAGENTS USED**

- Ortho-phosphoric acid (concentrated)

- Sodium hydroxide pellets

- Silver nitrate, 0.01N standard volumetric solution

- Potassium iodide (KI) solution (50g/L)

- Ammonia solution (6N)

- Distilled water

# APPENDIX 2

## HYDROGEN CYANIDE CONTENT

Hydrocyanic acid expressed in milligram of HCN per 100g of sample is equal to

$$\therefore 0.54 \; (V_0 - V_1) \times \frac{250}{120} \times \frac{100}{m} = \frac{112.5 \; (V_0 - V_1)}{m}$$

**FOR SAMPLE (C)**

$$C_1 = \frac{112.5 \; (16.10 - 12.45)}{20} \text{ where blank} = 12.45 \text{ for all conditions}$$
$$HCN = 20.53 ml/g$$

$$\text{For } C_2 : HCN = \frac{112.5 \; (16.58 - 12.45)}{20} = 23.23 ml/g$$
$$HCN = 23.23 ml/g$$

**FOR SAMPLE P**

$$\text{For } P_1 \; HCN = \frac{112.5 \; (15.40 - 12.45)}{20} = 16.59 ml/g$$
$$HCN = 16.59 ml/g$$

$$\text{For } P_2: HCN = \frac{112.5 \; (15.86 - 12.45)}{20} = 19.18 ml/g$$

HCN = 19.18ml/g

**FOR SAMPLE S**

For $S_1$ HCN = $\dfrac{112.5\ (18.45 - 12.45)}{20}$ = 33.75ml/g

HCN = 33.75ml/g

For $S_2$ HCN = $\dfrac{112.5\ (19.00 - 12.45)}{20}$ = 36.84ml/g

HCN = 36.84ml/g

Where;

H = Mass of test portion (20)g

$V_0$ = Volume in ml of 0.01N silver nitrate solution used for the sample determination

$V_1$ = Volume in ml of 0.01N silver nitrate solution used for the blank test.

# APPENDIX 3

## CORRESPONDING ETHYL CARBAMATE CONTENT

## FOR SAMPLE C

$\therefore 1.346 + 0.01 (X) = Y$

$C_1 = 1.346 + 0.01 (20.53) = 1.55 mg/l$

$C_2 = 1.346 + 0.01 (23.23) = 1.57 mg/l$

## FOR SAMPLE P

$\therefore 1.346 + 0.01 (X) = Y$

$P_1 = 1.346 + 0.01 (16.594) = 1.5119 mg/l$

$P_2 = 1.346 + 0.01 (19.18) = 1.53 mg/l$

## FOR SAMPLE S

$\therefore S_1 = 1.346 + 0.01 (33.75) = 1.68 mg/l$

$S_2 = 1.346 + 0.01 (36.84) = 1.71 mg/l$

From table 3.1, $y = a + bx$ (regression equation)

Since the values for a and b are meant to be determined using the pair of simultaneous equations.

$$\Sigma y = na + b \ \Sigma x \qquad \text{equation (1)}$$

$$\Sigma xy = a \ \Sigma x + b \ \Sigma x^2 \qquad \text{equation (2)}$$

Using the first values of the measurable cyanide (mg/100g) and it's corresponding ethyl carbamate (mg/100g)

If $x = 0.5$, then $Y = 1.00$, and also from table 3.1, $\Sigma x = 27.25$,

$Vy = 7.00$, $\Sigma xy = 47.82$, $\Sigma x^2 = 988.15$, $\Sigma y^2 = 11.24$

Substituting $\Sigma y = 1.00$, $n = 5$, $\Sigma x = 27.25$ in equation (1)

$$\therefore 47.82 = 27.25a + 988.15b \qquad \text{equation (4)}$$

Therefore solving equation (3) and equation (4) simultaneously to obtain values of a and b.

$$7 = 5a + 27.25b \qquad (3)$$

$$47.82 = 27.25a + 988.15b \qquad (4)$$

Multiplying equation (3) by 27.25 and equation (4) by 5

$$\therefore \ 7 \ (27.25) = 5(27.25)a + 27.25(27.25)b \qquad (5)$$

$$47.82(5) = 27.25(5)a + 988.15(5)b \qquad (6)$$

190.75 = 136.25a + 742.56b                    (5)

239.10 = 136.25a + 490.75b                    (6)

Subtracting equation (6) from equation (5) to eliminate a

48.35 = 0 + 4198.19b

48.35 = 4198.19b

b = $\dfrac{48.35}{4198.19}$ = 0.01

b = 0.01

Therefore substituting b = 0.01 in equation (3) so that a will be obtained as follows

7 = 5a + 27.25 (0.01)

7 = 5a + 0.2725

7 − 0.2725 = 5a

6.7275 = 5a

a = $\dfrac{6.7275}{5}$ = 1.3455 ≈ 1.346

a = 1.346

Therefore a = 1.346 and b = 0.01

# APPENDIX 4

## REAGENT PREPARATION

### SODIUM HYDROXIDE

5g of sodium hydroxide (NaOH) pellet was dissolved in small quantity of distilled water and then diluted to 200ml mark.

### SILVER NITRATE

0.85g of silver nitrate $AgNO_3$, was dissolved in small quantity of distilled water and then made up to 500ml mark and stored in a coloured bottle.

### POTASSIUM IODIDE

0.498g of potassium iodide (KI) salt was dissolved in water and the made to come up 300ml mark on the flask using distilled water.

## LIST OF ABBREVIATIONS AND ACRONYMS

AFC - Scientific Panel on Food Additives, Flavourings, Processing Aids and Materials in Contact with Food of EFSA

BMD - Benchmark dose

CAS - Chemical Abstracts Service

CN- Cyanide

CODEX - The Codex Alimentarius Commission was created in 1963 by FAO and WHO to develop food standards, guidelines and related texts such as codes of practice under the Joint FAO/WHO Food Standards Programme.

CONTAM - Scientific Panel on Contaminants in the Food chain of EFSA

CYPs - Cytochrome P-450 enzymes

DATEX - EFSA´s unit on data collection and exposure

DNA - Deoxyribonucleic acid

EC - European Commission

EU - European Union

EFSA - European Food Safety Authority

FAO - Food and Agriculture Organisation

GC - Gas chromatography

HCN - Hydrocyanic acid or hydrogen cyanide

IARC - International Agency for Research on Cancer

JECFA - Joint FAO/WHO Expert Committee for Food Additives

KCN - Potassium cyanide

LB - Lower bound

LCBO - Liquor Control Board of Ontario

LD50 - The lethal dose 50 (LD 50) is defined as the amount of a substance which causes the death of 50% of a group of test animals

LOD - Limit of detection

LOQ - Limit of quantification

MOE - The margin of exposure (MOE) is defined as the reference point on the dose-response curve (usually based on animal experiments in the absence of human data) divided by the estimated intake by humans

MRM - Multi Reaction Monitoring

MS - Mass spectrometry

NaCN - Sodium cyanide

ND - Not detected, that is results at or below the limit of detection or the limit of quantification for the method used

UB - Upper bound

Volume vs. mass Ethyl carbamate concentrations expressed as µg/L have been assumed to be the same expressed as µg/kg. However, the specific gravity for individual products will vary from just below 0.9 to just above 1.1.

WHO - World Health Organization.

Printed by Books on Demand GmbH, Norderstedt / Germany